```
J                          503475
391                         12.95
Mos
Moss
How clothes are made
```

DATE DUE			

GREAT RIVER REGIONAL LIBRARY

St. Cloud, Minnesota 56301

COSTUMES AND CLOTHES

How Clothes Are Made

Sue Crawford

MARSHALL CAVENDISH
New York . London . Toronto . Sydney

Reference Edition Published 1989

Published by Marshall Cavendish Corporation
147 West Merrick Road
Freeport, Long Island
N.Y. 11520

©Marshall Cavendish Limited 1989
©Wayland Publishing Ltd 1988

Editor: Deborah Elliott

Reference edition produced by DPM Services Limited

Consultant Editor: Maggi McCormick
Art Editor: Graham Beehag

All rights reserved. No part of this book may be reproduced or utilized in any form or by any means electronic or mechanical, including photocopying, recording, or by an information storage and retrieval system, without permission from the copyright holders.

Printed in Italy by G. Canale C.S.p.A., Turin
Bound in Italy by L.E.G.O. S.p.A., Vicenza

Library of Congress Cataloging-in-Publication Data

Moss, Miriam.
How clothes are made / Miriam Moss.
    p. cm. – (Costumes and Clothes)
  Bibliography: p.
  Includes index.
  Summary: Provides a history of the manufactoring of clothing.
  ISBN 0-86307-981-4
  ISBN 0-86307-980-6 (Set)
  1. Clothing trade–History–Juvenile literature. [1. Clothing trade– History.] I. Title. II. Series: Moss, Miriam. Costumes and clothes.
TT497.M69 1989
687–dc19   88-30215
                                 CIP
                                 AC

Contents

Chapter 1 — **Raw Materials**
- Natural fibers — 4
- Synthetic fibers — 6

Chapter 2 — **From Fiber to Fabric**
- Hand spinning — 8
- Machine spinning — 8
- Weaving — 10

Chapter 3 — **Design**
- Woven designs — 12
- Stitched designs — 13
- Finishing fabric — 14
- Designing clothes — 16
- The fashion industry — 17

Chapter 4 — **Knitting and Sewing**
- Knitting — 18
- Sewing — 22

Chapter 5 — **From Fabrics to Clothes**
- Handmade clothes — 24
- Factory-made clothes — 25
- How hats and shoes are made — 28

Glossary — 20
Books to read — 31
Index — 32

Chapter 1

Raw Materials

Natural fibers

For centuries, people have made yarn and clothing from natural fibers. Cotton and flax are vegetable fibers: they come from plants. Wool and silk come from animals.

Cotton

The cotton plant needs a hot, moist climate, so it is grown in countries such as Egypt, other parts of Africa, India, parts of North and South America, and China. The U.S. is the largest exporter of cotton in the world.

About three months after the cotton bushes have been planted, their seed pods, called bolls, ripen and burst open. The bolls are picked, and the long, downy fibers found inside are separated from the seeds at a **gin** and compressed into tight bales. The bales are sent to cotton mills all over the world, first to be spun into yarn and then woven to make a strong, comfortable fabric.

Flax

The stalks of the flax plant have been used to make yarn since at least 5,000 BC. Once the flax stems have been harvested, they are rippled, or combed by spikes, to strip off the leaves and seeds. Then the stalks are soaked to remove the outer part and the inner woody pith in a process called **retting**. Next, the flax is dried and passed through the revolving wooden beaters of a **scutching** machine. This separates and combs the flax fibers so they can be spun into yarn and then made into a fabric called linen.

Linen is a more **absorbent** fabric than cotton, and the fibers are smooth and not at all fluffy. It is so hardwearing that linen sheets which were used to wrap Egyptian mummies thousands of years ago have been found intact. Linen crushes easily and can split if it is washed incorrectly, so it is mainly used to make clothes which do not have to be cleaned very often.

A cotton picker at work on a plantation in Australia.

Above This mummy has been wrapped in linen since 200 BC — that's almost 2,200 years ago!

Below Spinning silk in a factory in China.

Wool

The coats of some animals also produce excellent fibers for making into yarn. Sheep's wool has been used for making fabric since prehistoric times. Furry animals such as alpaca goats, rabbits, and llamas also produce hair which spins into yarn well.

In the past, wool was probably plucked off the sheep by hand, but now it is sheared off in one piece, called a fleece. Long fibers are used for worsted yarn, which will be made into the best quality smooth fabric. After cleaning, the fibers are combed by a **carding** machine. Then they are put into a **gilling** machine which draws them out into thin slivers. Next, six slivers are wound by a machine into a single thread called a **roving**, ready for spinning. Shorter wool fibers are used to make hairy yarns and fabric.

Silk

Silk is produced by a type of caterpillar called a silkworm. To turn into moths, silkworms spin cocoons around themselves by squirting a sticky liquid out of glands in their bodies through two tiny holes, or **spinnerets**. As it dries, this thin stream hardens into a strand, or **filament**, of silk up to 1,500 yards long.

As the cocoons are unwound, several filaments are twisted together to make threads tough enough for weaving. Between five and eight cocoons yield about 300 yards of a thread called tram yarn.

Synthetic fibers

At the end of the nineteenth century, there was a shortage of silk. Scientists discovered that artificial (not natural) fibers could be made in the same way as a silkworm makes a strand of silk. They pumped a sticky mixture called viscose through tiny jets, or spinnerets, into a tank of acid where it hardened into one long thread. From this thread, a silky fabric called rayon could be woven.

Since then, many other artificial man-made fibers have been produced by treating natural and vegetable sources like wood pulp, milk, and cotton waste with chemicals. Wood pulp and cotton waste are used to make acetate and rayon. Milk protein produces fibrolane. Man-made fibers now account for over 20 percent of the world's **textiles**.

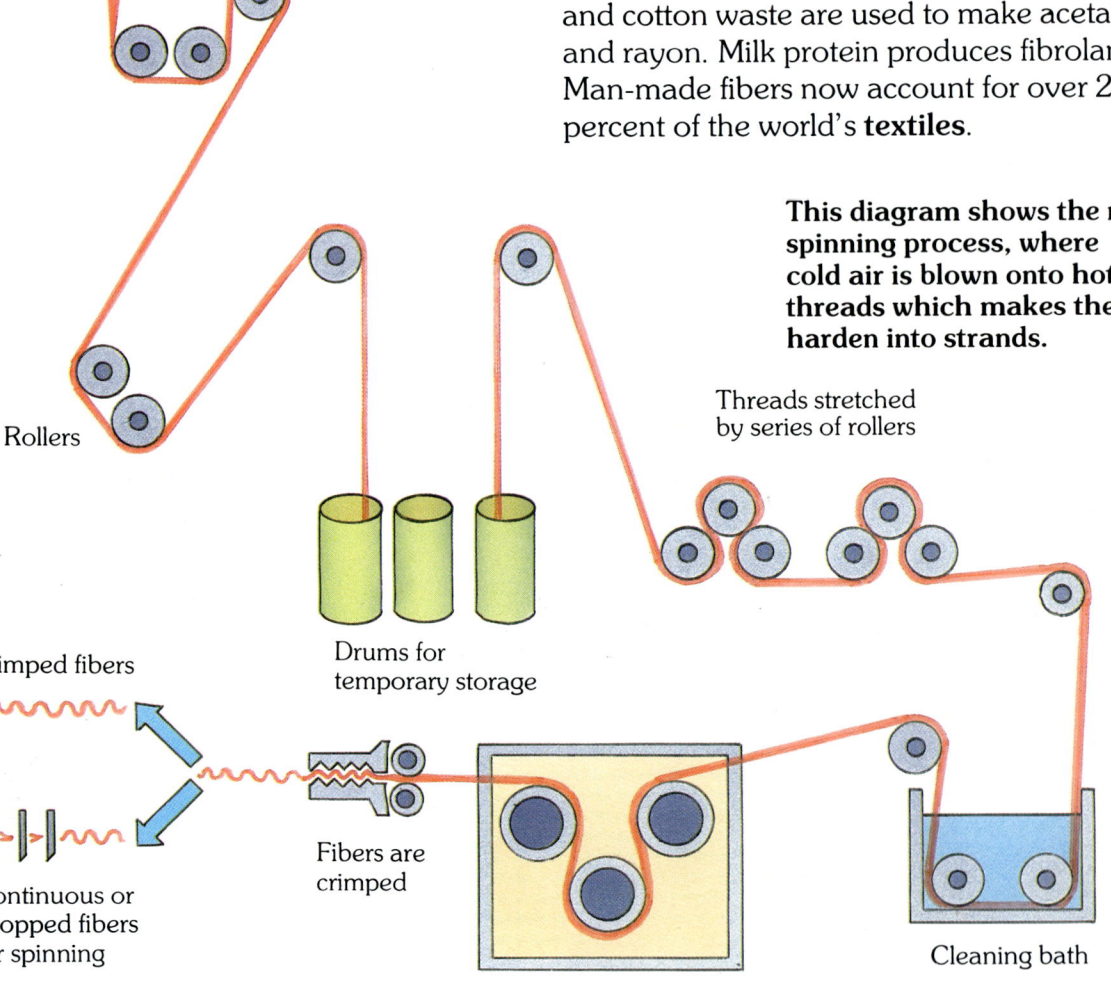

This diagram shows the melt spinning process, where cold air is blown onto hot threads which makes them harden into strands.

Synthetic fibers

Air, coal, oil, and water are used to make substances called **polymers,** which are treated with chemicals to produce synthetic (man-made) fibers. Well-known synthetic fibers are: polyamide fibers such as nylon, which is very elastic and is useful for making stockings; polyester fibers such as Dacron® which do not stretch and keep their shape well; acrylic fibers such as Orlon® which is soft and warm; and elastic fibers such as lycra, which is used for swimwear since it dries quickly.

Making fabric from synthetic fibers

Man-made fibers can be made into fabrics by knitting, weaving, or **bonding**. If the fibers are left as threads, they can be woven, like silk, into shiny fabrics. To make fabrics with more **texture**, the fibers may be cut into short lengths, like wool and cotton, and spun before weaving or knitting. Some synthetic fibers melt and stick together when they are heated. They can then be made into fabrics by using the bonding process.

In the bonding process, the threads are spread out evenly in all directions to form a web. When this web is heated, the fibers melt and stick together to form a sheet of fabric such as vilene. Vilene looks like paper, but is hard to tear. It is useful for putting between other fabrics to stiffen and strengthen them.

These Australian schoolchildren are enjoying their summer holidays on Palm Beach in Sydney. Their swimsuits are made from a synthetic material called lycra. Lycra is used to make swimwear because it dries quickly.

Chapter 2
From Fiber to Fabric

This woman in Ecuador is spinning wool by hand using a weighted stick called a spindle.

Hand spinning

Cotton, flax, and wool are all composed of many short fibers. In order to bind them into one thread long enough to weave or knit into fabric, the fibers must be spun into what is called **staple** yarn.

In some parts of the world, spinning is done by hand. Having cleaned the cotton or wool and combed it to take out the tangles, the spinner carefully draws out some fibers and twists them together. He or she then fixes them to a weighted stick called a spindle. The spindle is swung one way to help twist the yarn and then the other way to wind up the yarn until it is full.

During the thirteenth century, large spinning wheels were invented in Europe. With them, spinners were able to produce twice as much yarn as before, so they were widely used in factories for the next two hundred years.

By the sixteenth century, smaller "flying" wheels had been designed. A seated spinner could work them by using a foot treadle. Flying wheels wound the yarn onto the bobbin as it was spun. Although this was less tiring, it was not used in factories because the small wheels made the process slower.

Machine spinning

During the eighteenth and nineteenth centuries, hand-operated spinning wheels were replaced in many countries by power-driven machines. Nowadays, there is a big demand for cotton, the most widely-used fiber. Most countries that sell cotton to other countries use powerful machines to clean, comb, and spin raw cotton into yarn.

Mule spinning involves first stretching the fibers, then twisting them together; and finally winding them onto a bobbin. Ring spinning, where a machine is used to twist, spin, and wind the yarn in one single process, is faster, but it produces much coarser yarn. The latest ring spinning machines can carry 500 bobbins, which turn around 12,000 times each minute.

Rotor spinning is a much better method than either of these because there are no bobbins to be changed. Instead, the fibers are fed into the top of a moving container which flings them to the center and twists them together. They then pass out of the other end in one long string. Rotor machines can produce more than 120 yards of yarn in one minute.

This woman is spinning worsted yarn in a textile factory in Bradford, West Yorkshire, England.

Weaving

Weaving is the oldest method of turning all types of yarn into fabric. The earliest weaving was probably a form of darning. A number of threads, called the **warp**, were weighted and hung from the branch of a tree or stretched out and held in place by a row of pegs. Then another thread, called the **weft**, was passed over one warp thread and under the next in one direction, then back under and over warp threads the other way, to make a width of fabric. A firm edge, which is called the **selvage**, forms along the length of the fabric as it is woven.

Right An easy-to-use upright loom which is still in use in some countries today.

Diagram of an early loom.

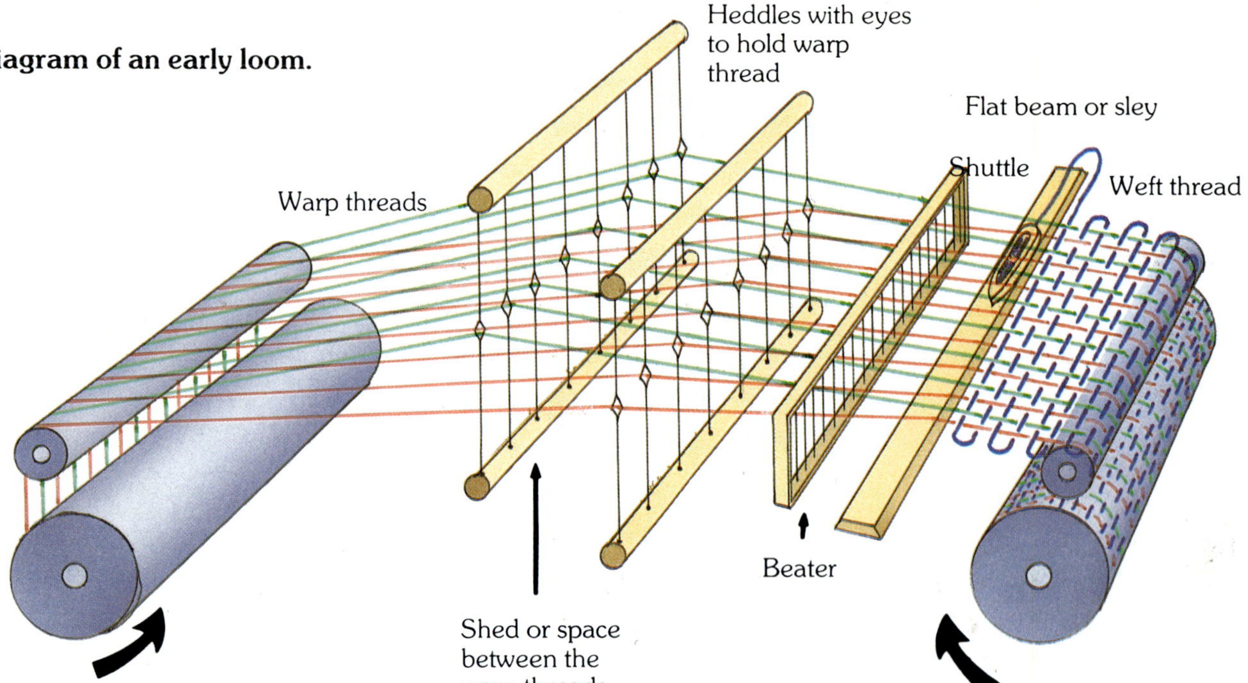

Heddles with eyes to hold warp thread

Flat beam or sley

Shuttle

Weft thread

Warp threads

Beater

Shed or space between the warp threads

Warp beam or roller

Cloth beam or roller

10

A woman weaving beautifully patterned carpets on a backstrap loom in India

The first looms were probably used in about 4,000 BC in the Middle East by **Neolithic** people. Looms made weaving much easier because a wooden heddle (the frame of the loom) raised every other warp thread to leave a clear path (or shed) for the weft thread to pass along. The weft thread was wound around a wooded shuttle and passed backwards and forwards across the lines of alternate threads to produce a plain weave.

At the start of the nineteenth century, a Frenchman named Joseph Jacquard thought of a way of using metal cards punched with holes to raise and lower different warp threads each time the weft thread passed across. Until then, the only way to vary the weave to make patterns with lots of detail was to lift each warp thread one by one using a "draw" perched on top of the loom.

During the nineteenth century, power-driven machines for weaving were invented and first used in the U.S. They had a self-threading shuttle and stopped automatically if the warp thread broke. Modern looms use air and water jets instead of a shuttle to send the weft across the warp. The latest machines can weave more than 400 yards of weft every minute.

Chapter 3 Design

Twill weave

Dogstooth weave

Herringbone

Satin weave

Woven designs

The illustrations on the left show herringbone weave, twill weave, satin weave, and dogstooth weave, which are four common woven designs. Leno is a fabric made by an open, lacy pattern of weaving. Dobby weaves are done in small figures of eight to make geometric patterns. The invention of the Jacquard loom in the nineteenth century (see page 11) led to the weaving of bigger patterns and more detailed designs called brocades.

To make fabrics like velvet and corduroy, extra threads are woven into the warp of a plain weave background to make a series of loops, which are later trimmed to make a nap. Terrycloth is a fabric which is made by leaving the loops uncut.

Right Embroidered lace is made by looping threads together with buttonhole stitch and darning other threads to make the pattern

Opposite A bobbin lace doily.

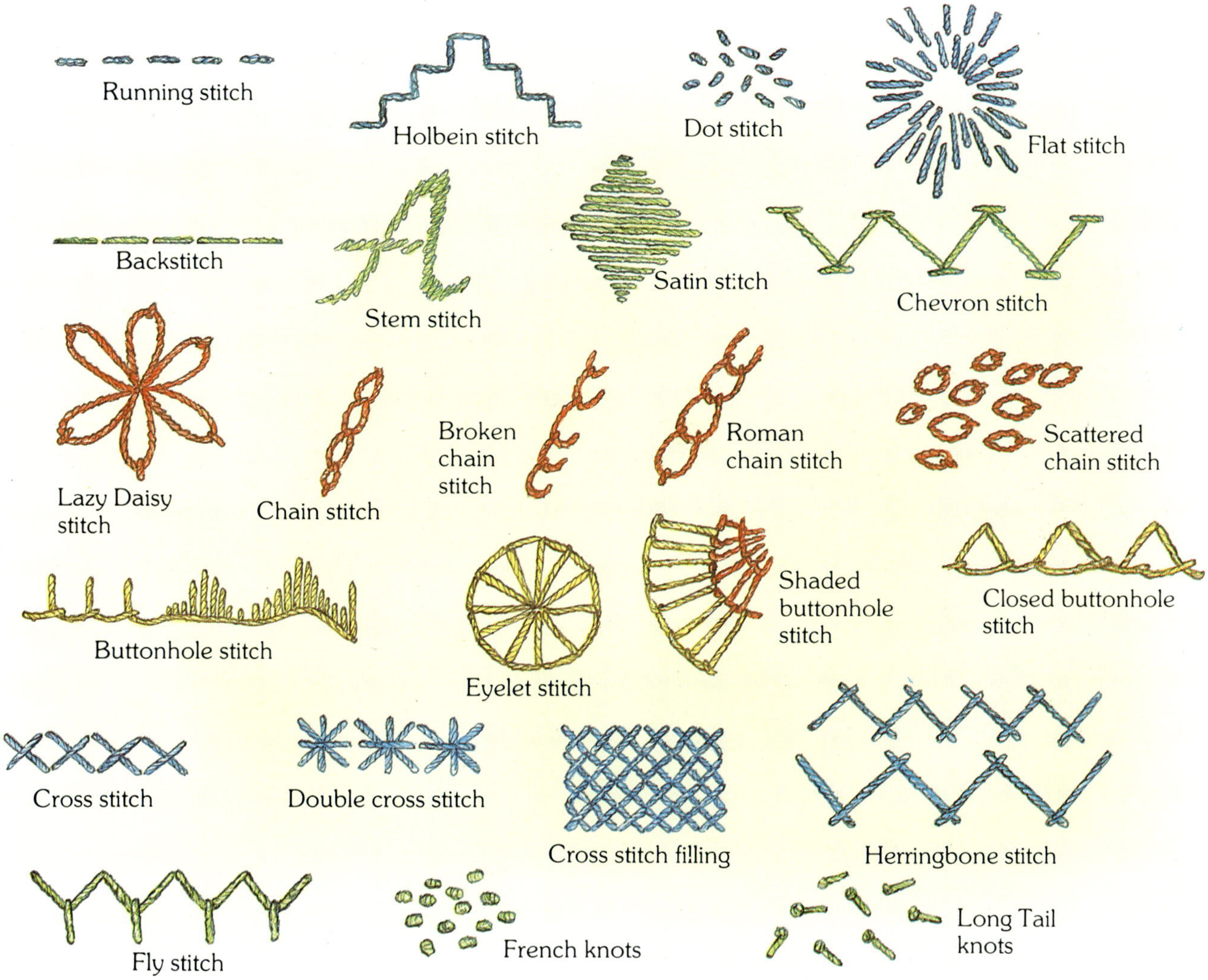

Running stitch

Holbein stitch

Dot stitch

Flat stitch

Backstitch

Stem stitch

Satin stitch

Chevron stitch

Lazy Daisy stitch

Chain stitch

Broken chain stitch

Roman chain stitch

Scattered chain stitch

Buttonhole stitch

Eyelet stitch

Shaded buttonhole stitch

Closed buttonhole stitch

Cross stitch

Double cross stitch

Cross stitch filling

Herringbone stitch

Fly stitch

French knots

Long Tail knots

Stitched designs

Patterns may be woven into the fabric by hand or machine. Tapestry goes back to the ancient Mesopotamians, who made fabrics by weaving colored yarns together. Nowadays, it usually describes building up a design by weaving colored threads into a web of plain backing fabric.

Embroidery is a way of decorating a plain piece of fabric by stitching colored threads onto it to form a pattern. Until this century, it was mainly done by hand. **Mass-produced** clothing can be embroidered on a Shiffli machine, which uses hundreds of needles guided by punch cards or a computer.

13

Finishing fabrics

An important part of designing fabric is to treat it in various ways. This can be done by dyeing, printing, or adding certain chemicals to protecct the fabric against such elements as heat and water. This is called "finishing the fabric."

Dyeing

Color is one of the strongest features of design, and dyes have been used to color textiles for more than 5,000 years. Sometimes, the yarn is dyed before it is made into cloth. In ancient Mesopotamia, the sheep were dyed before they had their fleece removed!

The earliest dyes were made from plants and insects, and many of these natural dyes are still in use today. The indigo plant yields a deep blue dye, and the saffron crocus a bright yellow one. To help the cloth retain natural dyes, it is soaked first in aluminium salts.

Nowadays, most dyes are made from **organic chemicals**. In 1856, the first synthetic dye, a reddish-blue color, was extracted from coal tar. Synthetic dyes tend to fade less than natural dyes.

Printing fabric

An early method of printing fabric was to stamp the pattern on with carved wooden blocks, one for each color. A faster, more recent method is roller printing, where the fabric is fed between a revolving drum and inked copper rollers which have the design cut into them.

For stretchy fabrics, where roller printing might blur the colours, transfer printing can be used. A colored design is printed onto special paper and then ironed onto the fabric. Screen printing is a way of stenciling designs onto fabric using a number of screens made from silk gauze stretched over wooden frames. Each screen lets a different part of the pattern through when colour is sponged over the top of it. Whichever sort of printing is used, the fabric must be steamed afterward to make sure that the colors last.

This Japanese woman is using a brush to varnish some beautifully patterned cloth.

Above This machine is drying fabric after it has been treated with chemicals to make it shrink-proof.

Above The fabric passes through a basin of colored dye as it winds from one roller to another.

In roller printing, fabric passes through a series of inked rollers. The bottom roller has a pattern engraved on it which is transferred onto the fabric.

Chemical finishes

Fabrics can be given all kinds of special finishes. They are important to the design because they affect the texture and the way the cloth can be used. Fabrics for outdoor wear are often made rainproof by oiling them with paraffin and metal salts, or with silicone. Fabric used for nightwear must be treated with **flame-retardant** chemicals to slow down the speed at which it burns. Fabrics can also be treated to make them mothproof or to prevent them from shrinking, staining, or creasing.

Designing clothes

Since people first wore clothing, every society has produced its own fabrics and fastenings, and decorated them by painting and embroidering them or using dyes to color them. People's needs for the climate and work they do have also affected the shape and texture of their clothes.

The Greeks and Romans draped lengths of cloth around the body as clothing. This useful design can still be seen today in the Indian sari. The Greek chiton was made from two pieces of fabric which were sewn halfway up the sides and fastened with clasps at the shoulders.

The Romans added a new design when they cut pieces away from the fabric to make the first tunic with sleeves. Tunics have been worn in many times and places. In medieval Europe, tunics were cut out in patchwork fashions, which means any fabric cut away to give the garment shape was sewn on somewhere else, to avoid waste.

Designing clothes purely for fashionable purposes began when fabric was cut and shaped to draw attention to different parts of the body. Until very recently, only wealthy people could afford fashionable clothes.

Gujarati women in India wearing saris.

The fashion industry

During the seventeenth century, fashions began to change more quickly, but it was not until the invention of sewing machines at the turn of this century, and paper patterns in 1917, that new styles of clothing were able to spread quickly and become widely available. Today, the average person sees many fashion changes in a lifetime.

In the twentieth century, the growth of international magazines which deal with fashion has spread the designs and ideas of a few famous designers, so that names like Chanel and Dior are linked with a certain style. Until the 1950s, Paris was the major center for new styles, but since then London, Rome, Milan, Tokyo and New York have each been important in starting new trends.

The mass production of ready-to-wear garments and the manufacture of many different, cheaper man-made fibers has allowed people in wealthy countries to buy fashionable clothes they can afford. Manufacturers only need to make a sketch of a design from a major fashion house for their designers to copy in order to produce patterns for a piece of clothing that looks new, individual, and fun — even though it may well be a re-mix of ideas from the past. The basic shape used today to make T-shirts differs little from the early Roman tunic.

Above A model wearing one of the season's new designs at the Paris Fashion Show. Fashion shows give people a glimpse of new styles.

Fashion designers usually make very rough sketches of their ideas, to see what they look like, before making the finished garment.

Chapter 4
Knitting and Sewing

Knitting

Knitting is an ancient way of making fabric. Pieces of fabric have been found which were knitted by the Egyptians. Knitting is formed by loops of yarn, linked together in rows. Unlike weaving, knitting allows you to use many different types of stitch in one piece of fabric, and to turn yarn into clothing with little or no sewing.

Most knitted clothes you buy from stores are made on factory knitting machines, but many people still hand-knit clothes at home. Find out on pages 20-21 how to knit a few basic stitches and make a scarf.

Right These boys in Ethiopia are wearing attractively designed sweaters which were hand-knitted in their area.

Knitting machines

The first knitting machine was invented in England in 1589. A long, springy hook, called a bearded needle, opened and closed to make each stitch. During the **Industrial Revolution**, power-driven "flat" knitting machines were built for use in factories.

Many knitting machines are now fitted with latch needles, invented in 1849. They work with a simple up and down movement which makes them very fast. Some machines still use bearded needles, however, because they can knit shaped garments and finer stitches.

Circular knitting machines, invented during the Industrial Revolution, turn continuously to knit tubular fabric. Since the 1960s, most nylon stockings and pantyhose have been knitted by this process.

Skirt fabric is often made on warp machines. Warp machines knit vertically (upright), by feeding two yarns to each needle, which is faster than having to go back and forth like the flat machines. In one minute, a warp machine with 2,500 needles can knit 1,200 rows of stitches, or three million knitted loops!

A knitting machine in a factory.

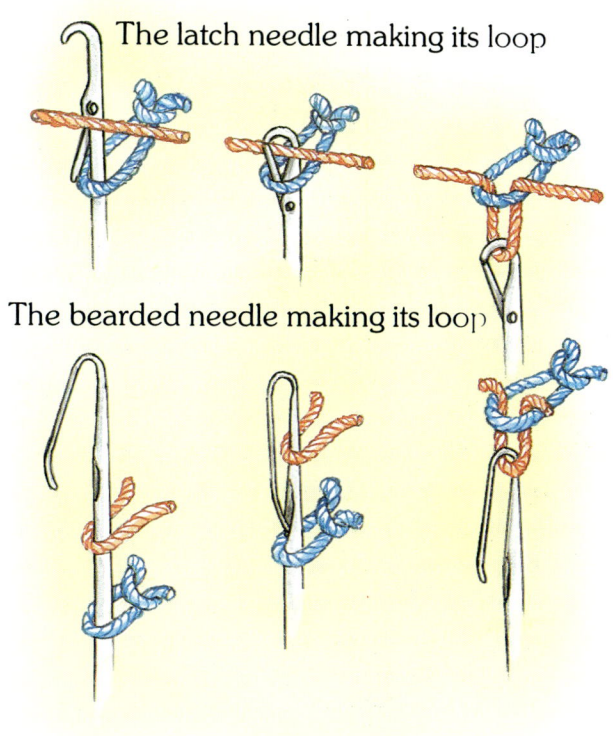

The latch needle making its loop

The bearded needle making its loop

Hand knitting

The quickest way to learn to knit is to ask an experienced knitter to show you how to begin. Knitting patterns tell you which yarn and what size of needles to use. Two or more fibers — such as wool and acrylic — are often blended to give the advantages of both. The label around the yarn tells you its color and weight. The **ply** number means how many strands of fiber there are in the yarn.

How to knit a scarf

To knit the scarf below, you will need a pair of number 6 needles, and 6 ounces of wool or acrylic knitting worsted yarn.

Casting on

1. Cast on 40 stitches; knot the yarn over one needle and draw a loop through it to hook on to the same needle.
2. Hold this needle in your left hand and pull a new loop through the space between your first two stitches to hook on to make the third stitch.
3. It is easiest to do this "hooking through" with the other needle held in your right hand.
4. Continue to pull loops of yarn between each new stitch and the previous one, and hook them onto the needle until you have a row of 40 stitches.

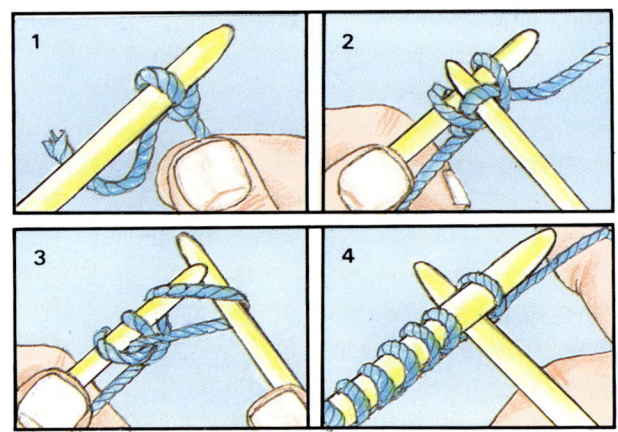

Knitting

1. Push your right needle into the first stitch on your left needle, and wind the yarn between the two needles.
2. Pull the loop of yarn through with your right needle, keeping the tips of the needles close together so the new stitch cannot slip off.
3. Carefully slide the old stitch off the left needle.
4. Carry on to the end of the row, by which time you will have transferred all the stitches from the left to the right needle.

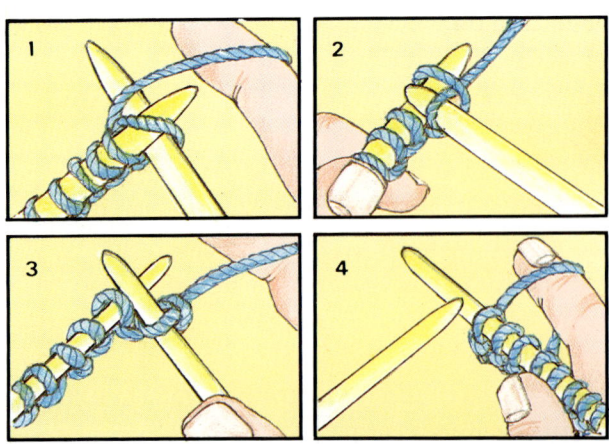

Garter Stitch

1. Move the full needle to your left hand, you will be ready to start the next row.
2. If you make your second row the same as your first, you will be doing garter stitch.
3. When this knitting is finished, it looks the same on both sides.

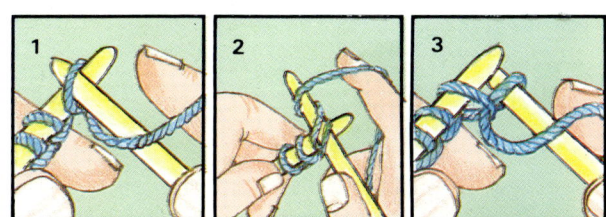

Purl Stitch

1. Purl stitches are very like knit stitches, but you begin with the yarn in front of your right needle.
2. Put the needle into the stitch toward you instead of away from you and wind the yarn around the right needle.
3. Pull the loop through and slip off the old stitch.

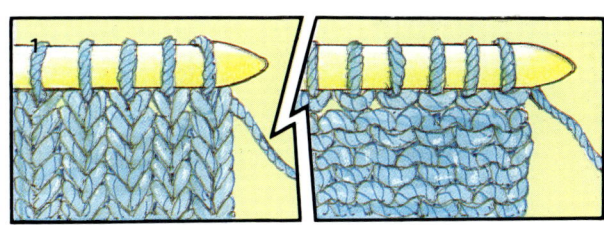

Stockinette Stitch

1. If you alternately knit one row and purl one row, you will make stockinette stitch, which looks smooth on the front and like garter stitch on the back. To make your scarf, continue with stockinette stitch until the knitting is about a yard long. Then finish by binding.

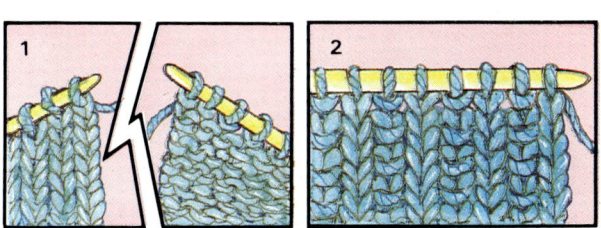

Ribbing Stitch

1. You can use knit and purl to knit other designs as well. By changing from a knit stitch to a purl on every stitch, you can make either rib or moss stitch.
2. To change from a knit stitch to a purl, or back again, you must remember to put your yarn around the needle so it is in the right place to begin that stitch. If you line up your knit and purl stitches, you will get a ribbed effect.

Binding off

1. When you have finished what you are knitting, you need to make an edge along the fabric which will not ravel.
2. Begin by knitting the first two stitches together.
3. Then, with your left needle, hook the first stitch over the second. Carry on in this way along the row until you are left with one stitch.
4. Break off the yarn and thread it through the loop; fasten it off securely.

21

Sewing

Sewing can be done by hand or machine. In hand sewing, two pieces of fabric are joined by threading a length of thread in and out through the fabric with a steel needle. Before steel was invented, needles were made from bone.

Sewing machines

Industrial sewing machines were invented in the U.S. in the 1850s. By about 1879, people were beginning to use smaller versions in their homes. Early machines were operated by a handle or foot treadle, but in factories they were gradually replaced by more efficient, power-driven machines. The latest electronic home sewing machines can do both decorative and practical stitches.

Making clothes in a factory in the Philippines

A sewing machine

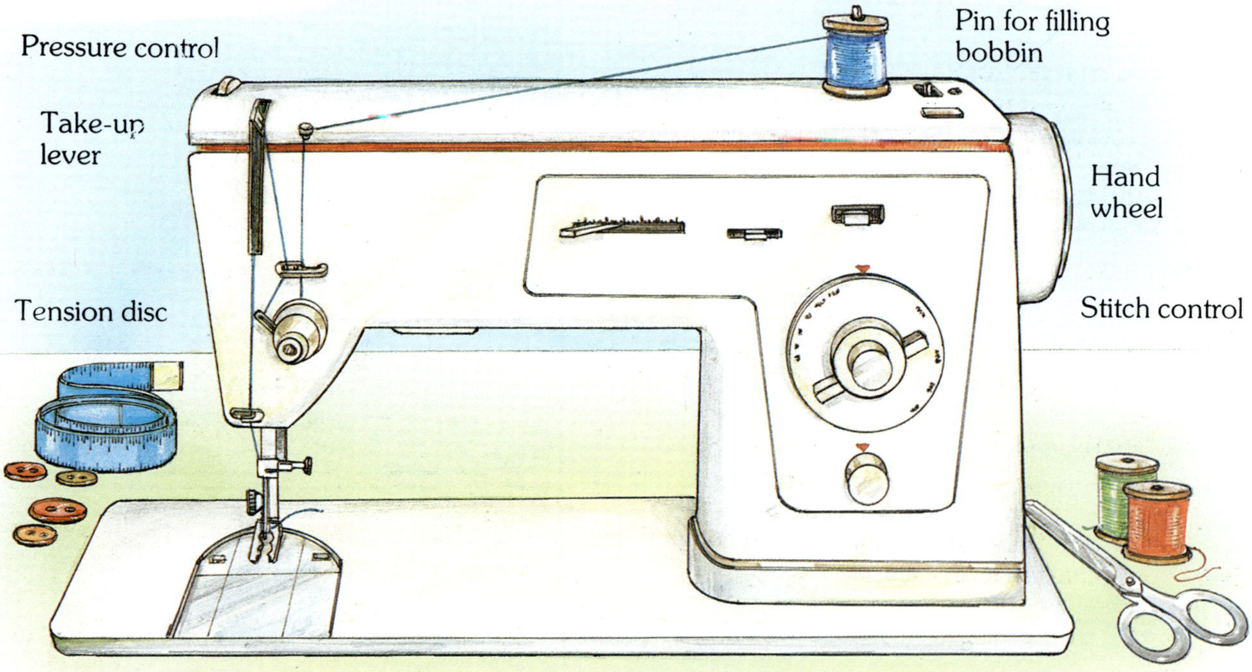

- Pressure control
- Take-up lever
- Tension disc
- Spool pin
- Pin for filling bobbin
- Hand wheel
- Stitch control

How to make a cotton top

To make a cotton top, you will need ¾yd of 36in 100% cotton material, one spool of matching thread, and some sewing needles. Remember to be very careful when using needles.

1. Make a pattern on dressmaker's graph paper (right) and cut it out. Fold your material in half; then pin the pattern to your material. Cut out around the shape of the pattern.

2. Pin the front and back together along the sides and shoulder. Make sure the right sides of the material are facing each other. Using large basting stitches (1 inch apart), sew the pieces together and remove the pins.

3. Sew the shoulders and sides together, using very short, neat stitches. If this stitching does not seem very secure, make another double row of short stitches.

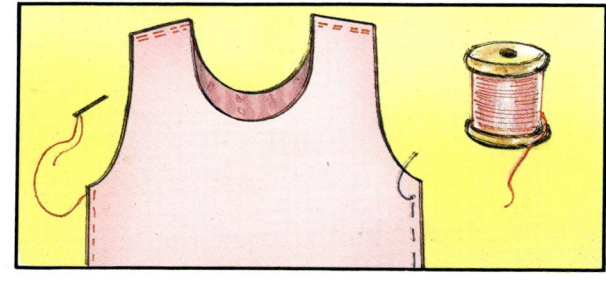

4. Take out the basting stitches. Using a warm iron, press the seams. (Ask a parent or older person to help you, as irons can be dangerous).

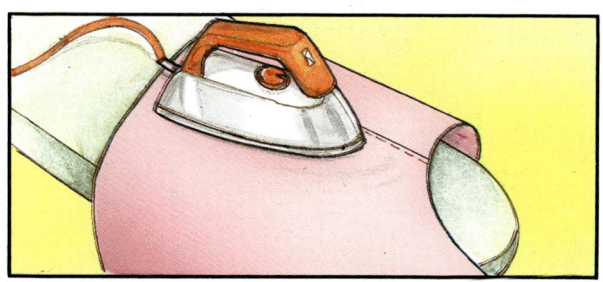

5. Turn in ¼ inch around the neck and armholes. Baste and press. Turn in another ⅜ inch to cover the raw edges. Baste and press this in place.

6. Again using short, neat stitches, sew around the neck and sleeve openings, about ¼ inch from the inner folded edge.

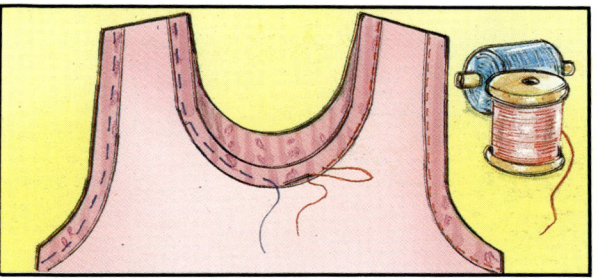

Chapter 5
From Fabrics to Clothes

Handmade clothes

Until two hundred years ago, spinning, weaving, knitting, and making clothes were mostly done by people at home or working in small businesses. In these "cottage industries" groups of people worked together to make clothes using skills passed down from their parents and grandparents. Today, however, there are many big factories which can produce many different kinds of clothes very quickly and cheaply.

Almost every country that produces clothes now uses mass production and the latest machinery to manufacture garments which can be sold all over the world. However, certain clothes are still skillfully made by hand or in small family businesses or community workshops around the world. Harris tweed is woven in this way by crofters who live in the far north of Scotland.

This man is using a mechanical knife to cut several layers of cloth at one time.

Above All material and finished garments have to be carefully checked for mistakes.

Above This machine is making holes in the material which will later be used as buttonholes

Factory-made clothes

Until the 1980s, the clothing industry still relied heavily on "outworkers," who are people who work in their own homes doing intricate jobs such as sewing zippers into factory-made jeans. Nowadays, many skilled and unskilled tasks are done automatically by machine. The latest zipping machines can insert sixty zippers in an hour!

In the factory, the pattern must first be laid on the fabric so the garment can be cut out with as little waste as possible. This is especially tricky if the fabric is patterned. After cutting, the pieces must be stitched together. Next, trimmings and fastenings such as buttons are added. Finally, the garment has to be freed of creases and pressed.

Automation has completely changed that part of the textile industry that manufactures millions of identical garments, such as sportswear and denim jeans. Computerized turntables turn the fabric around to insure that the pattern is laid with the least waste. The machines automatically cut up to 100 thicknesses at once.

The latest robotic sewing machines join up the garment automatically, they can even fit sleeves into coats. Finally, unwanted creases, which once had to be ironed out by hand, are removed much more quickly by passing the garment through a steam tunnel.

The flow chart on the next two pages explains the manufacturing processes involved in making clothes in a factory.

25

How clothes are made in a factory

1. Pattern making

In large companies, specially trained pattern technicians work closely with the designers to make sure that the pattern is exactly right.

From the designer's sketches, detailed drawings, with seams and measurements specified, are made.

A checklist is made of every piece that goes into the garment for the costing department. If the cost is too high, the garment has to be replanned.

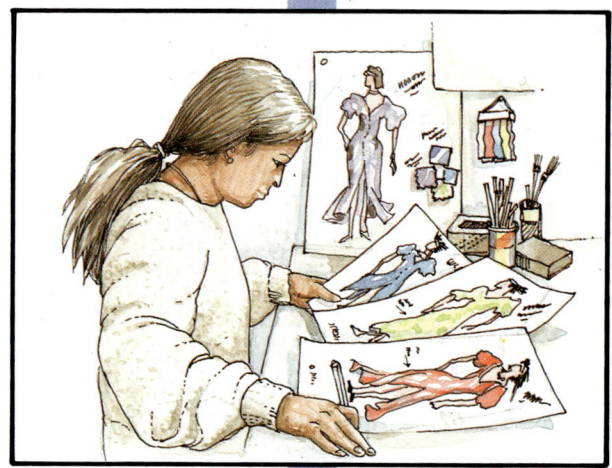

2. Lay planning

The designer or pattern technician passes the pattern pieces, all in different sizes, to the lay planner. All the pattern pieces have to be laid onto the fabric in the most economical way.

Mini computers are used to design the pattern pieces, grade them into different sizes, plan the most effective way of using materials, and produce a full-size paper pattern for manual cutting.

3. Cutting

The fabric is spread out on long tables in up to 150 layers by a "spreading" machine. Each layer is checked for flaws. The pattern pieces are laid on top and then cut.

There are many different types of mechanized cutting — straight knife, band knife, die cutting, computerized knife cutting, and computerized laser cutting.

4. Putting it all together

A conveyor belt carries the cut-out pieces to the person whose job it is to match them so they are ready for the sewing machinist.

Clothes are assembled sectionally. Each machinist concentrates on sewing one part of the garment on a specially designed machine.

Some factories have huge industrial sewing machines which can take the material, turn in the seams, sew it in place, and put the finished pieces back into the production line for the next operation.

5. Pressing

The work is pressed at various stages during production (this is "underpressing") and at the end of production comes the final pressing. The garments are laid on a huge flat table which is completely covered by a large flat press.

6. Quality Control

Work is examined for quality at various stages during production. There is always a final inspection which is done just after the final pressing.

27

How hats and shoes are made

Millinery

Hats have been worn since ancient times. Early hats, worn for warmth, were made of wool and looked like wigs. Later, hats made from plaited straw were worn to keep off the sun in country districts of Europe and Asia.

Making fashionable hats is called **millinery**. Hats are made on wooden skull blocks which come in different sizes. To find your size measure around your head from the top of your forehead, passing under the bulge at the back of your head. The basic shape of a hat is called "the crown." The diagrams below show how a crown is made or "blocked" from a special strawlike material called spartre.

Once the crown is made, any sort of covering can be added to make an interesting hat. Hats for special occasions can be made from velvet, fur, satin, or lace. A brim made from stiffened spartre and wire can be attached to the crown, along with ribbons, flowers, or other decorations, to finish it off.

1. Measure your head from your forehead to the back of your head.

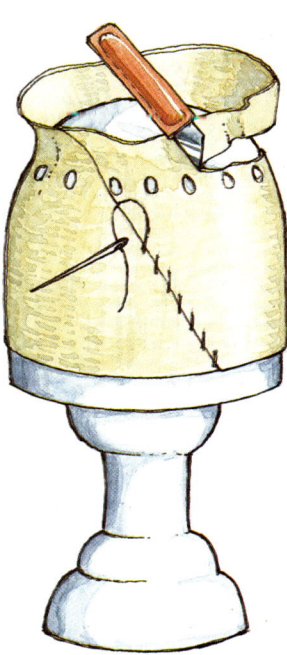

2. Material is placed on block.

3. Material is cut to correct size.

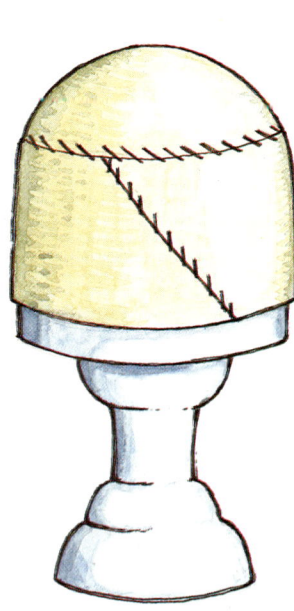

4. Material is stitched.

How shoes are made in a factory

1. Shoes are made in batches of 12. Uppers and vamps are made from foam backed nylon. A cutter cuts through the nylon using a hydraulic press.

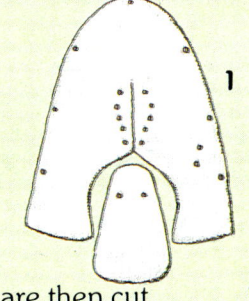

2. The suede leather trimmings are then cut. There are five different pieces for each shoe, and each has to be cut separately.

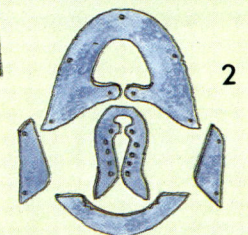

3. Each pair of vamps is fitted into a metal frame called a pallet. The suede pieces are positioned on top, and the pallet is put under an automatic sewing machine. The computer-controlled machine stitches the trimmings and vamp together. The tongues are then stitched together and attached to the vamp.

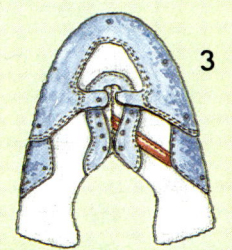

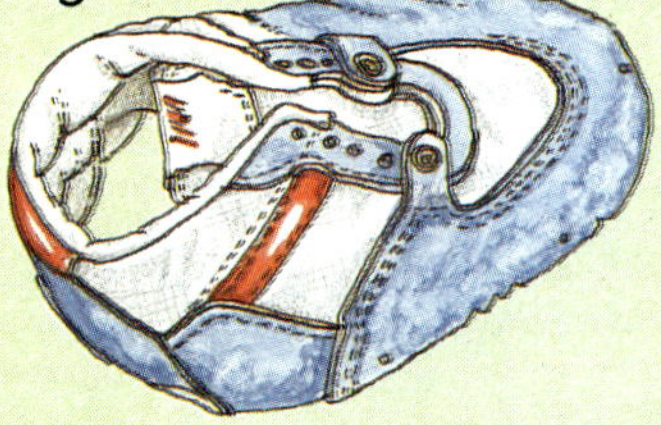

4. The heel lining is sewn onto the back of the shoe. The shoes are then run through a blade which trims away any extra material.

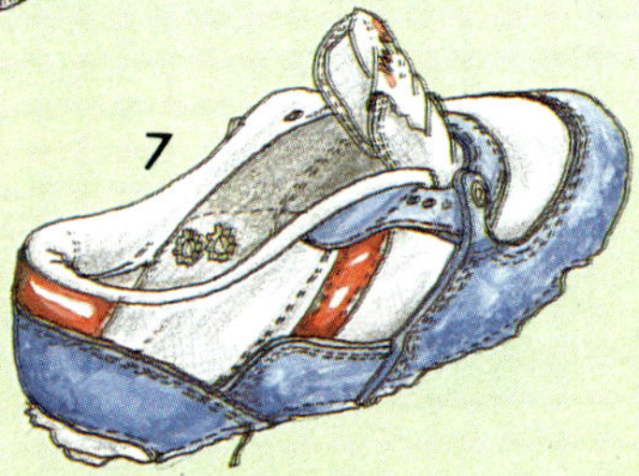

5. A machine punches lacing holes and eyelets in the shoes.

6. A canvas sock is sewn onto the bottom of each upper. This gives the shoe its shape.

7. The toes are overlocked — a machinist attaches two long strings onto each toe of the shoe. A long length of string is left on one side of the toe, which is then run through an overlocking machine. This binds the string securely to the toe and gives the toe its final, proper shape.

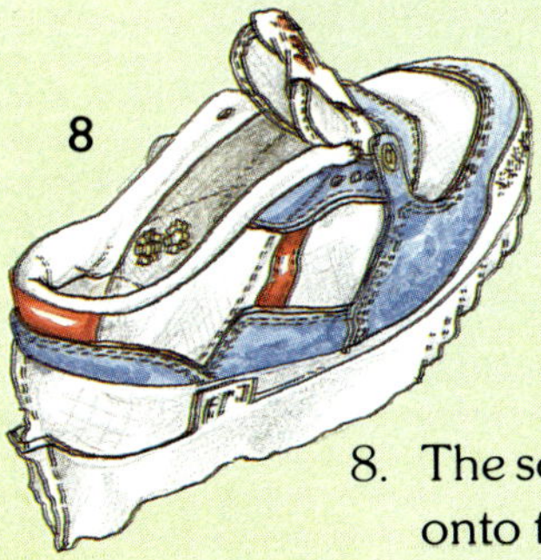

8. The soles and two side pieces are then molded onto the upper by an enormous molding machine.

9. The shoe is trimmed all the way around to give the molding a smooth edge.

10. The inside of each sole is brushed with glue and a foam sock is firmly stuck into place.

11. Shoes are inspected for quality.

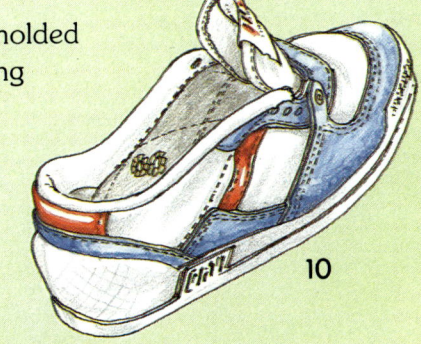

Glossary

Absorbent Able to soak up water

Automation Using automatic machinery to do a job instead of wokers to do it by hand.

Bonding Turning fibers into fabric by spreading them out and melting them until they fuse together to make a web of fabric.

Carding Untangling and separating wool or cotton fibers for spinning by combing them between two carders (brushes covered with wire teeth). Spiky plants called teasels were originally used. Nowadays, carding is often done by machine.

Filament A continuous, unbroken strand of fiber. Silk is the only natural filament fiber, but most man-made fibers are produced as filaments.

Flame-retardant A fabric which has been treated for safety reasons, to make it slow to burn if it catches fire.

Gilling Drawing wool fibers out into thin slivers ready for spinning.

Gin A plant in which raw cotton is separated into fibers and seeds.

Industrial Revolution The time during the eighteenth and early nineteenth centuries when there was a huge growth in industry in Britain due to the increase in the use of machinery.

Mass-production Large numbers of identical products made by a series of mechanized processes.

Millinery Making and selling women's hats and fashion accessories.

Neolithic The late Stone Age, when people first began to live in settled agricultural communities in the Middle East.

Organic chemicals An enormous family of chemicals, all derived from carbon.

Ply The number of strands twisted together in a length of yarn.

Polymers Synthetic fibers made by a chemical reaction which rearranges simple molecules into more complex structures that are strung together in long chains.

Retting Soaking flax stems to remove the hard outer part and inner woody pith.

Roving Several slivers of wool fiber loosely wound together ready for spinning.

Scutching Separating and combing natural flax fibers before spinning them into yarn.

Selvage The long, outside edge of a piece of woven cloth, which will not unravel.

Spinnerets Tiny holes under the mouth through which silkworms force out silk. Also, nozzle through which synthetic fibers are made.

Staple Fibers naturally found in short lengths, such as wool and cotton. Man-made fibers can be cut up into staple fibers.

Textiles Fabrics made from natural or synthetic fibers by bonding, weaving, felting, knotting, or knitting.

Texture The structure of woven fabrics, such as course, lumpy, rough, or smooth.

Warp Threads stretched lengthwise (or vertically) on a loom, so that weft threads can be woven under and over them at right angles.

Weft Threads woven back and forth across the warp threads to make cloth.

Books to Read

If you would like to find out more about how clothes, shoes and hats are made, you might like to read the following books:

Cloth: Inventions That Changed Our Lives by Elizabeth S. Smith (Walker & Co, 1985).
Clothes by Peter Curry (Price Stern, 1984).
Clothing (also in Spanish edition: *La Ropa*) by Judy O'Hare (Duggan Publications, 1985).
Costumes and Clothes by Jean Cooke (Franklin Watts, 1987).
Cotton by Millicent E. Selsam (William Morrow, 1982).
A Day in the Life of a Fashion Designer by Ann Hodgman (Troll Associates, 1987).
Folk and Festival Costumes of the World by R. Turner Wilcox (Scribner, 1965).
From Cotton to Pants by Ali Mitgutsch (Carolrhoda Books, 1981).
From Sheep to Scarf by Ali Mitgutsch (Carolrhoda Books, 1981).
The Sewing Machine by Beatrice Siegel (Walker & Co, 1984).
Silkworms by Sylvia A. Johnson (Lerner Publications, 1982).
Textile by Boy Scouts of America (BSA, 1972).
The Weaver's Gift by Kathryn Lasky (Warne & Co, 1981).

The publisher would like to thank the following for providing the pictures for this book: Camerapress 17 (top); National Institute of Cotton 11 (left), 15, 19 (top); Christine Osborne 7,18 (right); Julia Osorno 13 (top), 17 (bottom), 19 (bottom), 20-21, 22 (bottom), 23; Sefton Photo Library 24; Ronald Sheridan 5; Malcolm Walker 6, 10, 12 (left), 18 (left), 26-27, 28, 29; Wayland Picture Library 1, 3 (bottom), 22 (top), 25 (left); Zefa cover, 4, 5, 8, 9, 11 (right), 12 (right), 14, 16, 25 (right).

Index

Acrylic 7, 20
Animals 4, 5, 6
Automation 25

Bobbin 8, 9
Bonding 7

Carding machine 5
Chemicals, use of 7, 14, 15
Chiton (Greek) 16
Coloring 14, 16
Computerization 3, 25
Cotton 4, 6, 7, 8, 9
Courtelle 7
Crimplene 7
Crown 28

Designing clothes 16, 17
Designs 14
   stitched 13
   woven 12
Dyeing and dyes 14, 16

Embroidery 13, 16

Fabric 4, 5, 7, 8, 10, 12-16, 18, 19, 22, 25
   treatment of 15
Factories 8, 18, 19, 22
Factory-made clothes 25, 26, 27
Fashionable clothes 16, 17
Fibers 4, 5, 8, 9, 20
   elastomeric 7
   man-made 6, 7, 17
   natural 4, 6
   polyamide 7
   polyester 7
   synthetic 4, 7

Fibrolane 6
Filament 5, 6, 7
Finishing fabrics 14, 15
Flax 4, 8

Gilling machine 5
Gin 4

Handmade clothes 24
Hats 28

Knitting 7, 8, 18, 24
   hand 18, 20
   machine 18, 19
   needles 19, 20
   patterns 20

Linen 4
Loom 11, 12
   Jacquard 11, 12
   modern power-driven 11
Lycra 7

Machinery 8, 9, 11, 13, 17, 22, 24, 25
Making a cotton top 23
Making a scarf 18, 20-21
Mass production 13, 17, 24
Millinery 28

Nylon 7, 19

Paper patterns 17, 25
Pattern on fabric 11, 12, 13, 14
Plants 4, 14
Polymers 7
Printing material 14

Rayon 6
Retting 4
Roving 5

Sari 16
Scutching machine 4
Sewing 22
   by hand 18, 22
   by machine 17, 22, 25
Shiffli machine 13
Shoes 28, 29
Silk 4-7, 14
Silkworms 5, 6
Spindle 8
Spinnerets 5, 6
Spinning 4, 8, 9, 24
Steam tunnel 25

Tapestry 13
Terylene 7
Textiles 6, 14, 24, 25
Texture 7, 15, 16
Tunic 16, 17

U.S.A. 4, 11, 22

Velvet 12, 28
Vilene 7
Viscose 6

Warp 10, 11, 12, 19
Weaving 4, 5, 7, 8, 10-12, 13, 18, 24
Weft 10, 11
Wool 4, 5, 6, 7, 8, 20

Yarn 4, 5, 8, 9, 10, 13, 14, 18, 19, 20, 22